I0484339

WITH APPRECIATION TO:

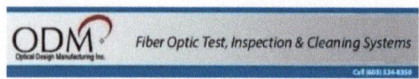

FOR TEST
SAMPLES
AND
TECHNICAL
SUPPORT

...and to many thousands of craftsmen,
hundreds of trainers, manufacturer's reps,
business friends/foes, and committee
participants who provided insights into
the many facets of fiber optic
deployment,   production and design.

2

INTRODUCTION:

There are many cleaning products. Some have been used for decades, centuries and millennia: others newly available. Since The Clean Air Act of the early 1990's, new solvent chemicals, including water-based cleaners, have changed the way (just about) everything is cleaned!   Over the years, wiping cloths have evolved from 100% cellulose (paper) and cotton to 100% synthetic fibers or combinations such as non-woven hydro-entangled materials.

Precision cleaning fiber optics has benefitted from this in some ways, and in others there is a sense of confusion.

<u>There are three recognized procedures to clean a fiber optic connection.</u>

a.) A "DRY" PROCESS

b.) A "WET-TO-DRY" PROCESS

c.) A "HYBRID" OR "COMBINATION" TECHNIQUE

THERE IS A 4TH PROCEDURE: it is "Blind-Cleaning"™ : the assumption that *any process or product* cleans *any* debris or contamination in a leap-of-faith <u>without</u> video inspection.

The principals in this book are applicable to all connection types. The intent of the work is to open a dialogue with you, designer, installer, or producer so there is a sense of clarity at the time when cleaning a fiber optic connection. It is often confusing, controversial, and ever so important to the future of fiber optic deployments as a competitive telecommunications media.

Thank you for purchasing this abridged edition of a study featuring the Optical Zonu® SFC Transceiver.
There is an additional version with another 60+ pages of the laboratory study with over 100 photographic comparisons of the cleaning processes and tools. It may be purchased from Amazon, Amazon-EU, on Kindle® or www.createspace.com/5120367.

A transceiver is the heart-line connection from the "outside/in" to the electronics of transmission equipment. Connected to a transceiver is a "jumper-side" connection.

If the "jumper" is not connected to a transducer, there 'back plane' connection must also be cleaned. There is also potential for 'alignment sleeve' contamination. All must be considered and properly cleaned: there are always 'two sides to every connection'.

"Best Practice Protocol" is to clean both sides every time a connection is "opened". Also, this is true with 'factory cleaned' connections. *Why? Too many potentials for contamination and the message to 'clean some and not others' is confusingly inappropriate.*

A fiber optic connection may be deployed in pristine-like data centers or in less than desirable conditions. *As is the case with cleaning anything, there are many variables.*

THESE INCLUDE:

1.) The connection type itself. The cleaning principals presented in this book will work on all connections, however, the products themselves may not.

2.) The deployment environment is an understated and often overlooked consideration.

An installation may be performed in a data center with controlled environment, or, in a snow storm along a right-of-way, on a flight line, or in an entertainment venue. *Each presents a different challenge.* In the ever-expanding world of fiber optic deployment there are many potentials of "dry debris", "fluidic contamination" and "combinations" of the two!

A critical first step to a successful cleaning result is the question the technician asks her/himself: *"what type of debris or contamination can I anticipate here today?".* Plan accordingly: likely, every site is different.

3.) The knowledge of the craftsman is a critical variable. Some are aware cleaning is important while others rely on simplicity and convenience. Others struggle to clean multiple times and finally "pass" what may not be best. Testing itself many be inaccurate when a removable contamination is mischaracterized as an irremovable 'artifact'.

4.) Not all cleaning products or procedures work the same way to remove the same type debris or contamination each and every time. Video inspection is critical.

5.) In this book we investigate the fiber optic end face. In another study we will focus on "FUSION SPLICE PREP". The two are different methods and procedures and should not be interchanged.

4

What Does It Mean?
"Future-Proof"

➢ The <u>process</u> that anticipates future developments

➢ Actions taken to minimize negative consequences.

➢ Seize opportunities.

➢ Change old perceptions into new realities

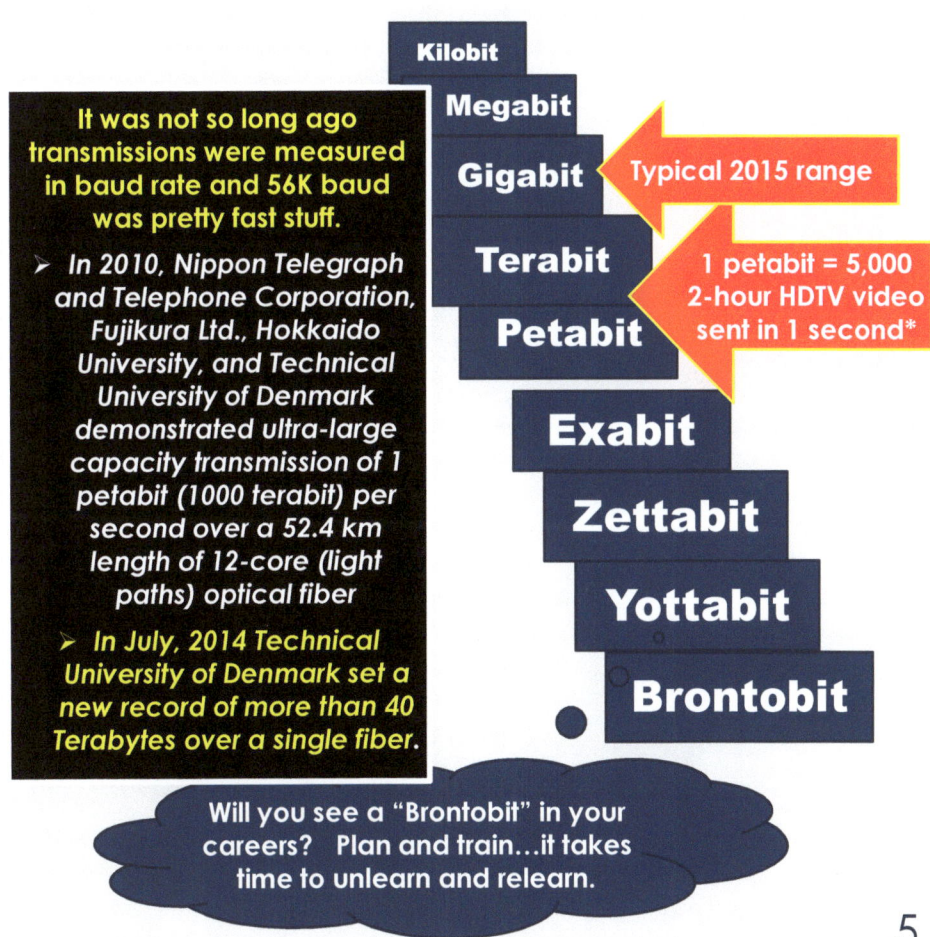

It was not so long ago transmissions were measured in baud rate and 56K baud was pretty fast stuff.

➢ In 2010, Nippon Telegraph and Telephone Corporation, Fujikura Ltd., Hokkaido University, and Technical University of Denmark demonstrated ultra-large capacity transmission of 1 petabit (1000 terabit) per second over a 52.4 km length of 12-core (light paths) optical fiber

➢ In July, 2014 Technical University of Denmark set a new record of more than 40 Terabytes over a single fiber.

Kilobit
Megabit
Gigabit
Terabit
Petabit
Exabit
Zettabit
Yottabit
Brontobit

Typical 2015 range

1 petabit = 5,000 2-hour HDTV video sent in 1 second*

Will you see a "Brontobit" in your careers? Plan and train…it takes time to unlearn and relearn.

In short:

- At the time of transmission, the fiber optic end face (and surrounding area) is **"the weakest link"**.

  - Multi-Mode or Single-Mode, it is Best Practice to clean each type every time the connection is 'opened' or 'installed'.

- **Why?**

  - Unlike copper, which likely never was cleaned or needed to be clean, it is essential to instill the need to precision clean fiber optics.

- **Why?** Unlikely there will be terabyte Ethernet transmissions over copper and...in this world of security concerns...fiber optics remains the first choice.*

  - End User demand for capacity and speed will increase...as it has from times of 'fax machines' and 'baud rates' from only a few years ago.

* Although, those 'pesky hackers' seem to be capable of just about anything!

DID YOU KNOW?

There are two instruments to view a fiber optic connection. One is a 'direct view' microscope and the other is a 'video inspection scope'.

For eye safety, a 'direct view' scope should <u>never</u> be used when there is any possibility an 'active' transmission could be inspected.

A light source is not a reliable instrument to assure the fiber optic end face is actually 'clean'.

There are various segments to the SFP-type…and any fiber optic connector type.

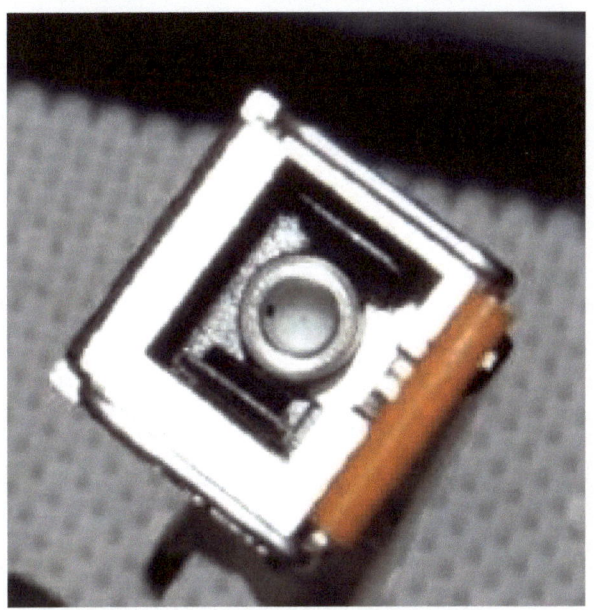

International Standards, test equipment and training places an emphasis on one small area:

There are various segments to the SFP and
any fiber optic connector type:

a.) The first and most common consideration
is the end face and the area of the "core"
or "stub".

There are various segments to SFC and
any fiber optic connector type:

b.) There are secondary areas. In this case
the "side wall" of the adapter.

If this outside diameter is not cleaned
debris can be transferred to the end face area.

On an 'adapter' for "SC" or "LC" connections, this
area is called the 'alignment sleeve'. There are
many recesses in an alignment sleeve where
contamination can 'reside' and are never seen.
This contamination can become present in the time
of post installation and test.

There are various segments to SFC and
any fiber optic connector type:

c.) There are secondary areas. In this case
the "inter-space" of the connector/adapter.

All surface areas must be considered: dry debris
or fluidic contamination can transfer.

While these areas may not be cleaned each and every
time, they can be a source of contamination if cleaning
and re-cleaning does not provide a pristine surface.

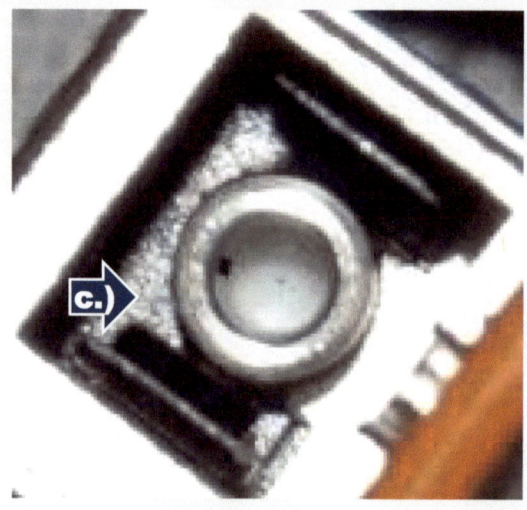

Author's Technical Note: For the purpose of
this work, I define "dry debris", "fluidic
contamination" and "combination soils" as
general categories

There is more than an "end face" even in Expanded Beam Connections

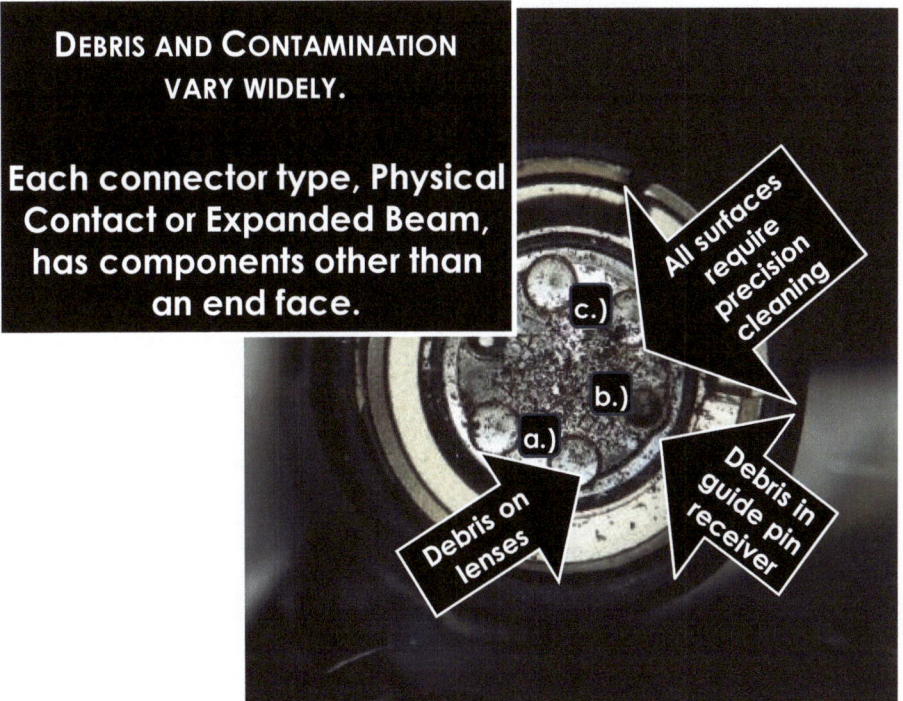

DEBRIS AND CONTAMINATION VARY WIDELY.

Each connector type, Physical Contact or Expanded Beam, has components other than an end face.

All surfaces require precision cleaning

c.)

b.)

a.)

Debris on lenses

Debris in guide pin receiver

DID YOU KNOW?

Precision cleaning fiber optics follows some very basic things... you already know! Dry debris tends to stay in place while fluidic contamination can move around. Both types can transfer one-side-to-the-other and throughout connector geometry.

There is more than an "fiber end face" to the MT® Type Connection

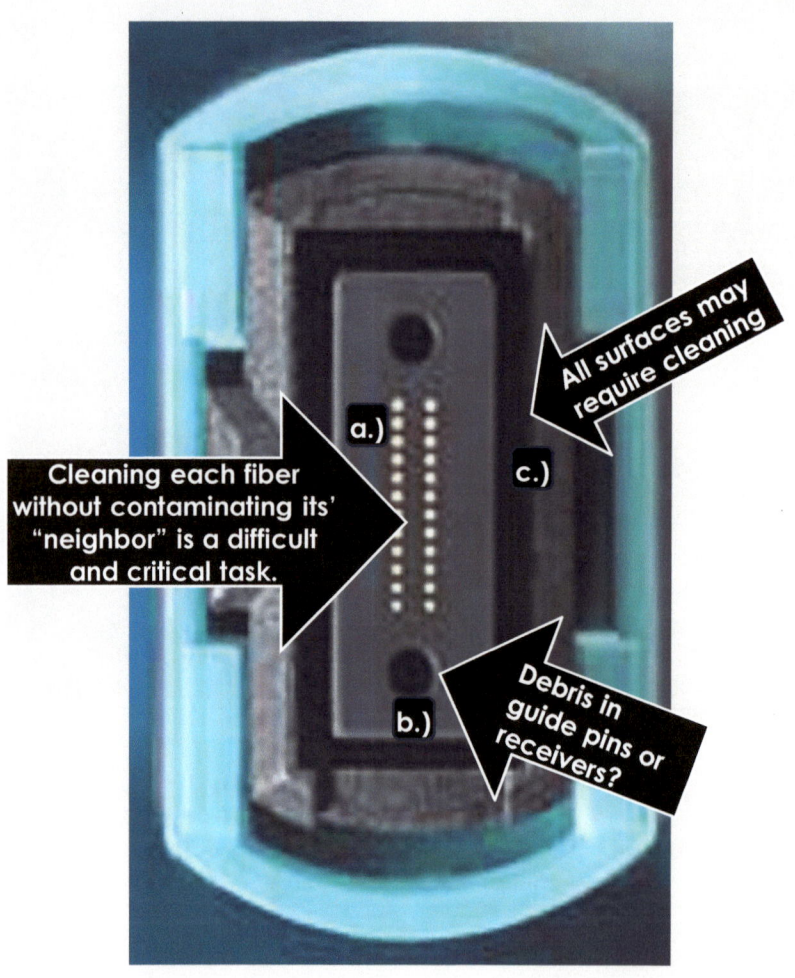

Cleaning each fiber without contaminating its' "neighbor" is a difficult and critical task.

a.)

c.)

All surfaces may require cleaning

b.)

Debris in guide pins or receivers?

There is more than an "end face" ...than just an 'end face'....

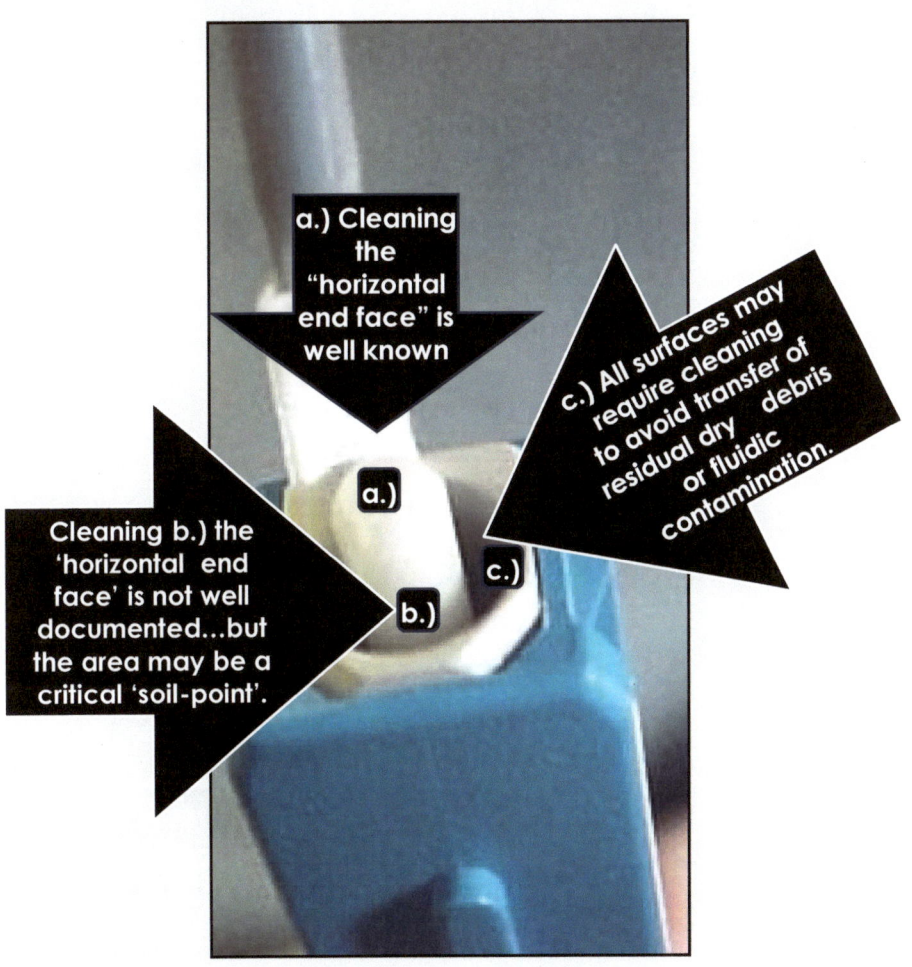

a.) Cleaning the "horizontal end face" is well known

c.) All surfaces may require cleaning to avoid transfer of residual dry debris or fluidic contamination.

Cleaning b.) the 'horizontal end face' is not well documented...but the area may be a critical 'soil-point'.

a.)

b.)

c.)

The 'end face' is the
'business end'
of the fiber cable and
signal transmission.

Is there is one cleaning
process for all connector
types and all Industry
segments?   Can one
product do it all?

This work will lead you to better
conclusions: the 'best practice'
for your business type, specific
connector and specific installation.

Cleaning the connection is based on principals starting from ancient times. There is little mystery about cleaning: the principals follow tenets dating to the Babylonians who (likely) invented soap, and, hydrocarbon cleaners used in the Arabian world many hundreds of years ago.

Debris and contamination are attracted to moisture: recall images of persons washing cloths along the Nile River, contemporary dishwashing, floor cleaning, automobile washing apparatus for your vehicle.

Few would take a dry cloth and remove sand or dust from a new car's painted surface!

...and in this case, there is a need to determine which "applications-specific procedure" is appropriate for your design, installation, or manufacturing operation.

A fiber optic end face is 'the Super-Car of communications transmission'!

How is a connection cleaned to "best practice"?

The "short answer":

DEPENDS ON THREE THINGS…

1.) Debris or contamination type.

2.) Is the connection video inspected?

3.) What cleaning tools are used?

- Remember, a light source and power meter will <u>not</u> indicate a clean connection.

- An OTDR <u>will</u> provide a reflectance or insertion loss measurement that can lead to the conclusion an end face is contaminated.

- A video inspection scope is the only reliable and practical way to assure the connection is 'clean'.

- As a safety measure, a 'direct view' scope must always be avoided where there is any possibility of an active transmission.

Why does the connection "need to be cleaned"?

The "short answers":

1.) "Dry Debris", "Fluidic Contamination" and "Combinations" can create reduce signal strength. This many mean a video transmission is pixilated or lost completely, upload and download rates distorted, and, customer satisfaction proportionately reduced!

2.) A soiled connection adds to an insertion loss budget that may include polish and splice loss among the factors.

3.) As fiber optic deployments continue to increase, technicians who never managed fiber, such as CoAx or Category trained craftsmen, did not have to clean a copper connection and *may not have the awareness of this basic need.*

4.) System and network designers also should be aware that precision cleaning affects performance.

17

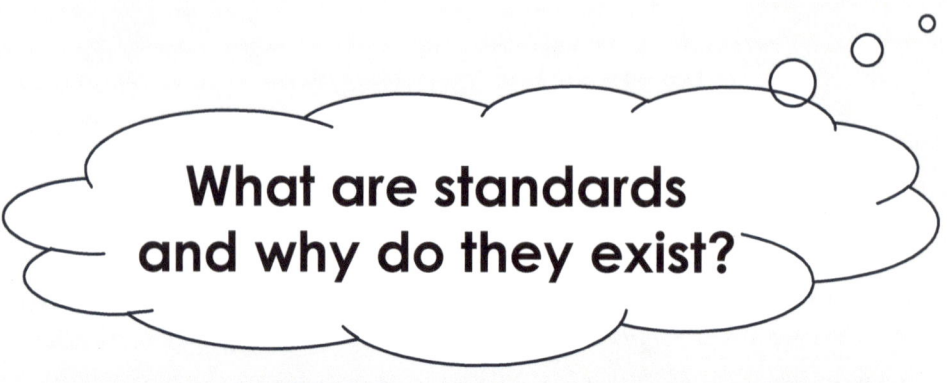

What are standards and why do they exist?

Honed by time…and some of the finest minds in our Industry.

Some Existing Standards:

- IEC61300-3-35: 2008 and update in 2018
- IECTR62627: 2008 and update in 2018
- TIA 455-240: 2008 and uncertain update
- Telcordia GR2923-Core: 2010 and update uncertain
- SAE AIR 6031: 2013 and update in progress

- ✓ **Standards 'set the pace'**
- ✓ **Technical advances outpace Standards.**
- ✓ **Good? Bad? Be aware…**

Most typically, standards run on a 5-10 year cycle

REGARDING FIBER OPTIC
TRANSMISSIONS, MOST ARE OUTDATED
AT THE TIME THEY ARE PUBLISHED.

WHY?

a.) Science of Transmission outpaces standard publication dates

b.) Investment in test and measurement follows the standards.

It may be time to think of standards as a 'base line', *minimum…or…*

Establish your own.

General Types of Debris and Contamination

Dry Type Debris	Fluidic Type Contamination	Combination Contamination
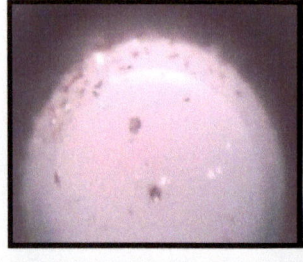		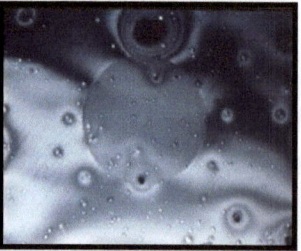
i.e.: Dust, Lint, Sand, Soils	i.e.: organic (cooking oil, engine oil, etc.), inorganic (salts) hand lotions. Water residues, silicones.	Combinations of dry and fluidic. Unidentifiable types

---

DID YOU KNOW?

There are particles of dry debris so small they can 'surface bond' to an end face. This surface bonded debris can dislodge 'post cleaning and inspection'. Precision cleaning considers all these aspects.

Please inquire about the White Paper study of surface bonding

Dry Debris and Fluidic Contamination are Three Dimensional.

Existing standards consider debris and contamination only in a two-dimensional diameter.

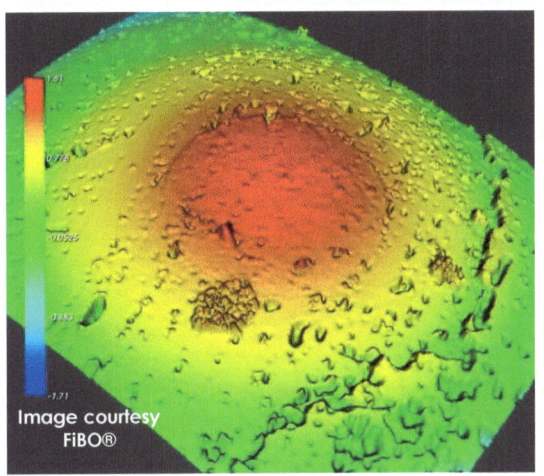

Image courtesy
FiBO®

INEFFECTIVE CLEANING CAN ALSO RESULT IN:

1. Static Field Contamination from the dry cleaning process*

2. A standoff between the jumper and backplane end face

3. Damage if the dry debris is harder than the ceramic ferrule or end face material

4. Fluidic recontamination outside the field of view can transfer one side to the other at time of insertion, post-cleaning and post-inspection

➢ In this interferometer reading, in some places, the height of this contaminant is greater than the diameter.

Please inquire about the study on Static Field Contamination

THE FIBER OPTIC END FACE IS <u>ALSO</u> A 3D STRUCTURE!

Existing standards in 2015 only consider an area of a radius of the core in a Zone System such as: Zone 1-2-3 (seen below).

Likewise, most video inspection "sees" (only) that limited area.

There are actually Five Zones in the 3-D Structure

<u>EXPLAINING "FIELD OF VIEW"</u>

a.) Existing standards are based on IEC 61300-3-35 which limits the viewing area to a Zone A-B-C-D which terminates at an area seen below as "Zone-3" or a 250-300 micron radius of the 'core': Zone-1

b.) A higher, more practical standard is the 4-5 Zone criteria which considers the complete end face as well as a wider range of possible contamination.

c.) Existing cleaning standards use a limited, easy to remove "dry debris" and/or "fluidic contamination". To date, there are no standards that include combinations of debris.

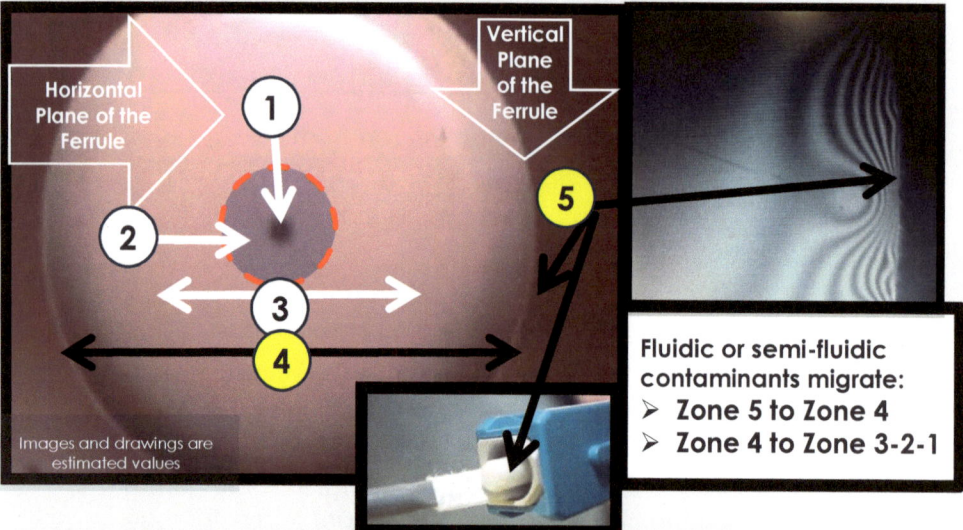

Horizontal Plane of the Ferrule

Vertical Plane of the Ferrule

Images and drawings are estimated values

Fluidic or semi-fluidic contaminants migrate:
- ➢ Zone 5 to Zone 4
- ➢ Zone 4 to Zone 3-2-1

Author's Technical Notes:
- Zone-1 is the core
- Zone-2 is the cladding
- Zone-3 is the area 250-300 micron radius of the core: typical 400x video inspection field-of-view,
- Zone-4 is the complete 'horizontal' end face
- Zone-5 is the 'vertical' end face.

22

Too much cleaning solvent can "flood" the connection...leaving excessive 'residue in areas that are never 'seen'.

Select precision fiber optic cleaners that remove the widest range of debris and contaminants.

**Not all are the "same".
Chose based on "performance".**

Not using a precision cleaner can result in static field attraction of additional dry debris.

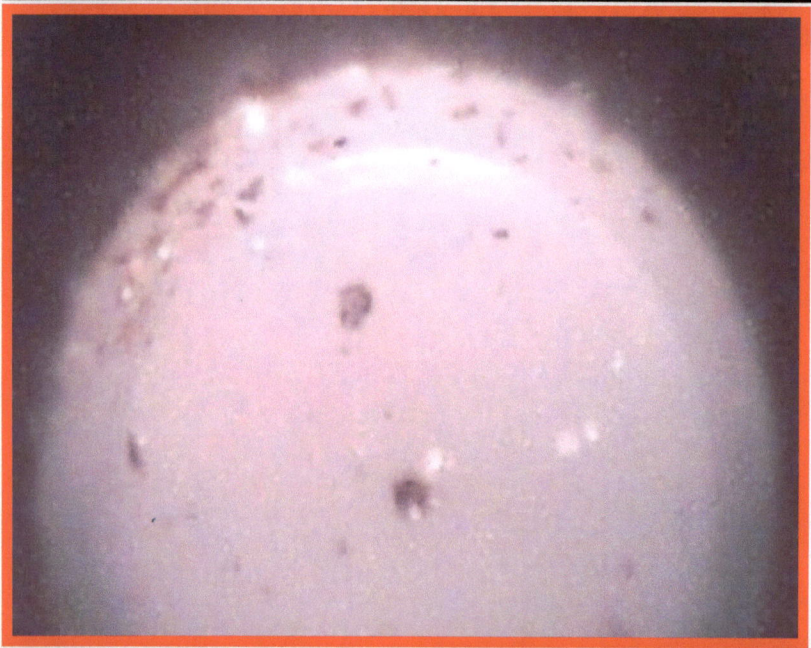

STATIC IS CONTROLLED IN TWO WAYS:

1.) "Conducted" by use of a grounding strap.

2.) "Dissipated" by use of a solvent

➢ No 'conductive path' can be created to manage static field contamination. Use of a precision solvent is the only option when static anticipated in areas of low relative humidity or temperature extremes.

➢ Static treatments or treated cleaning products can create residues in addition to the original debris or contamination.

Precision Cleaning a fiber optic connection is really a 'common sense' approach:

1. A fiber optic connection is, logically, three dimensions.

2. There is a "horizontal aspect surface" as well as a "vertical aspect surface.

3. With speeds and capacities of fiber optic transmissions ever increasing, precision cleaning means to properly clean by considering all potential sources of contamination

4. Dry Debris, Fluidic Contamination and 'Combinations' of the two logically are also three dimensions.

5. While "dry debris" may stay in place, this type of contamination also has height.
   - *An ineffective cleaning procedure can leave a 'removable' residue that may be mischaracterized by software as an 'un-removable artifact.'*

6. "Fluidic Contamination" may transfer not only from one end face surface to the other, but also within the connection geometry.

"Future-Proof" not only the installation, but also the craft of the technician, by understanding these matters as "Best Practice"

Coffee and Donuts...
...Pen and Paper
Soap and Water...
...Rum and cola
Wet and rain...
...Laugh and happy

Life is filled with many combinations:

> IN REGARD TO FIBER OPTIC DEPLOYMENTS, INSPECTION AND CLEANING ARE INEXTRICABLY TIED TOGETHER.

> Proper cleaning and proper inspection assures "Best Practice" deployments for all fiber optic installations

How is a connection cleaned?
What is "best practice"?

- Identify the components of the connection and assure you have the proper adapters to visually inspect.

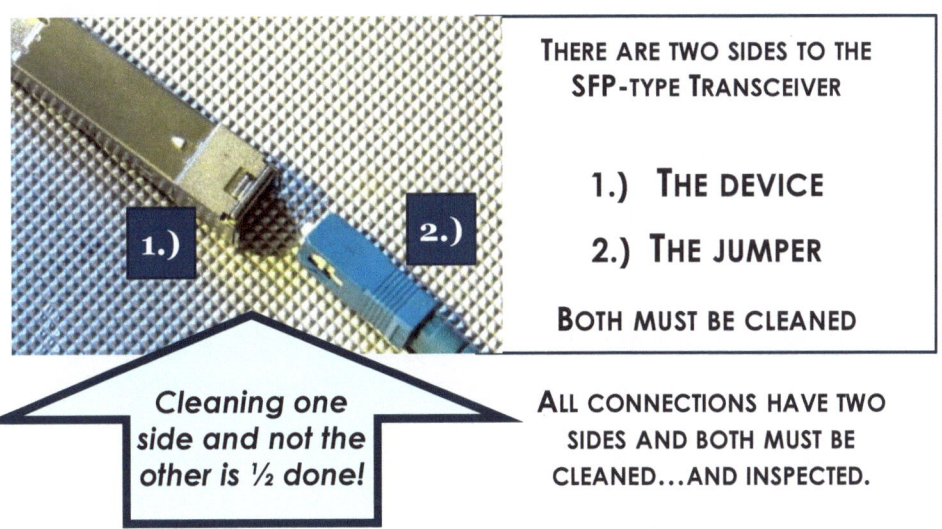

THERE ARE TWO SIDES TO THE
SFP-TYPE TRANSCEIVER

1.) THE DEVICE

2.) THE JUMPER

BOTH MUST BE CLEANED

Cleaning one side and not the other is ½ done!

ALL CONNECTIONS HAVE TWO
SIDES AND BOTH MUST BE
CLEANED...AND INSPECTED.

How is the connection cleaned to "best practice"?

The Importance of Video Inspection

Specific SFP Adapters are required to Video Inspect the transceiver 'stub'. You will need adapters for 2.5mm, 1.25mm both UPC and APC polish.

DID YOU KNOW?

A light source and power meter reading will not indicate if the surface is clean. Only video inspection can provide this information.

HOW IS THE CONNECTION CLEANED TO "BEST PRACTICE"?

BY CHANGING TIPS,
INSPECT THE 'JUMPER SIDE"

This 'field of view' video scope has high resolution to see debris and contamination details as well as more of the end face horizontal surface.

There are two sides to every fiber optic connection and both must be inspected and precision cleaned.

There is also an 'alignment sleeve' that can collect debris and contamination. It is not practical to inspect the alignment sleeve on most connection types. This means a specific cleaning process must do this task.

PROBE TOOLS ARE CONVENIENT AND FAST.

Their cleaning surfaces are either narrow thread-like material or wider tapes.

The narrow thread-like material cleans a smaller surface area; the tapes clean a larger surface as well as have the ability to 'accumulate' larger amounts of debris or contamination. Some of these tools will clean the alignment sleeve; others cannot.

Select and use these devices, as is the case with any of these products, based on the types of debris to be removed.

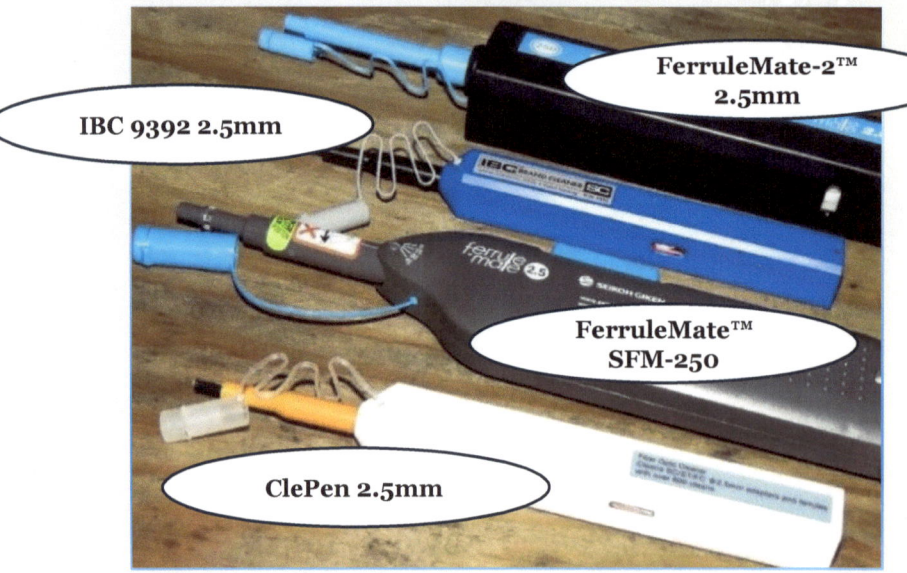

FerruleMate-2™ 2.5mm

IBC 9392 2.5mm

FerruleMate™ SFM-250

ClePen 2.5mm

How is the connection cleaned "best practice" using swab tools?

In this evaluation Sticklers®, NTT®, and ITW Chemtronics® 2.5mm swab tools are considered.

Precision Swab tools are used in demanding applications such as Class-1 operational clean rooms, wafer fabs and medical procedures.   Sometimes these tools are called "sticks" which can give the impression they have hard surfaces. They do not.

They are also appropriate for fiber optic back plane cleaning. Some can clean the alignment sleeve; others cannot. (FIGURE-1) Debris on the alignment sleeve can be passed to the   opposite end face.

• Select from 2.0/2.5mm, 1.25mm, or MT® types.

SWAB TOOLS™ HAVE TWO SURFACES (FIGURE-1)

a.) one that cleans the end face (black highlight)

b.) a separate surface that cleans the alignment sleeve (yellow highlight).

If the diameter of the swab is designed appropriately the alignment sleeve will be cleaned as well as the back plane end face/stub. Probes may not clean the alignment sleeve as the tapes and threads are too small to touch that surface.

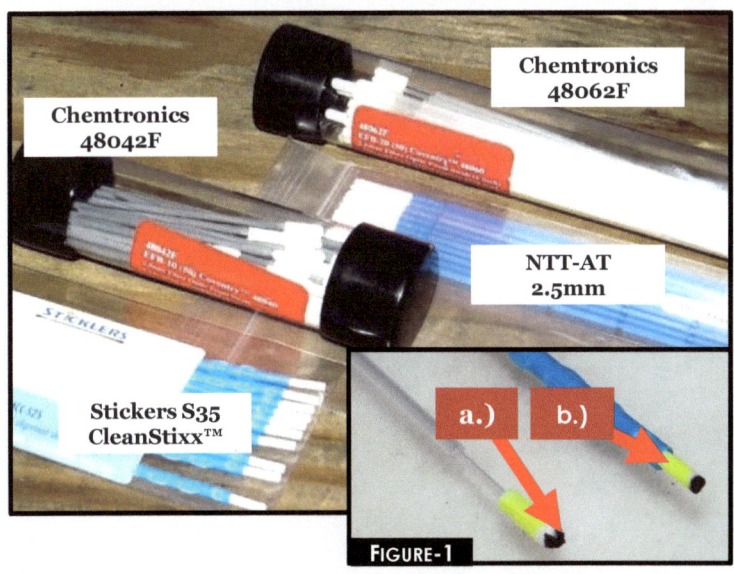

Chemtronics
48062F

Chemtronics
48042F

NTT-AT
2.5mm

Stickers S35
CleanStixx™

a.)   b.)

FIGURE-1

Cleaning Platforms™ and Reel Cleaners are used to clean the "jumper side" of the connection

---

THE LARGER SURFACE OF THE CLEANING PLATFORM™ HAS SEVERAL ADVANTAGES:

a.) Debris or contamination is moved away from the initial point of contact. *Swab tools and probes rotate the contamination in the same place. The small surface of a swab tool or probe may be overly-taxed to remove some debris or contaminant types.*

b.) The larger surface enables a perpendicular approach of the end face to the cleaning surface. This means that on an angled connection there is greater likelihood of first time cleaning without snagging on the bevel.

c.) In some instances, swab tools and probes will not clean the "back plane". In this case, the 'cleaning platform™ is used to clean by entering the equipment rack from the rear.

- A debt of gratitude should be offered to the "reel cleaner". It's invention in the late 1990's lead the way to understand the basic need to properly clean a connection. Reel cleaners are "convenient".

A FEW WORDS ABOUT CLEANING MATERIALS:

WIPING MATERIALS:

The materials in all tools are either 'woven' microfibers or 'non-woven' hydro entangled combinations of materials. 100% polyester can generate static field contamination.

- 100% cellulose (paper) is not appropriate nor is cotton.

  - There is no such thing as 'lint-free' cellulose (paper) or cotton materials.

  - A cleaning material must not contribute to contamination.

- Some cleaning materials work well on their own and others do not requiring a solvent or other component.

  - Among others, a decision factor is the type of debris or contamination to be removed.

  - As of this time, there is no ESD-wiper that would not impart impurities.

- Swab tools have materials that range from cleanroom-grade foam to sophisticated polyester and microfiber or hydro-entangled combinations. *All can be highly effective.*

FIBER OPTIC CLEANERS:

Alone, high-purity (99.9%) isopropyl alcohol is not effective

- a.) Its' hygroscopic nature draws moisture into itself causing dilution of cleaning ability

- b.) Even as a 99.9% pure chemical it is not as capable as the solvents tested to clean a wide range of debris and contamination.

- c.) IPA may be a component in a precision cleaner in ranges from ~4% to ~20%.

- Since the 1990's new generations of precision cleaners have evolved. Some are highly effective, others not so much! *Consider "performance" and balance against other trade-offs:*

  - HFE-7100 based

  - HFC based

  - Precision Hydrocarbons

  - 'Aqueous' Glycol Ether Cleaners

  - Others

INVESTIGATE MANUFACTURER'S CLAIMS.

33

WHAT ARE SOME PROCEDURES FOR AN EVALUATION?

Here are eight types of debris and contamination for an internal standard or product evaluation.

For the purposes of this study, a "dry-type" contamination is termed "debris", a "fluidic-type" is termed "contamination", and parts of both are termed a "combination-type".

1.) Arizona Road Dust

2.) Afghanistan Desert Dust

3.) 10w-40 SAE Engine Oil

4.) Jergens® Hand Lotion

5.) Wesson® Vegetable Oil

6.) ATD and Jergens® Lotion

7.) Afghan desert Dust and 10-w40 Engine Oil

8.) Wesson Oil and ATD

Author's Technical Note: Be creative! In 2014 I trained a series of technicians in (far) Western Kansas. Their "problem contamination" was dry dust from cattle pens and massive feed lots.

DID YOU KNOW?

Coarse sand from Huntington Beach is different from the powdery soft texture of the beach at Cancun! The same is true with Arizona Road Dust, which is coarse and easier to remove than the powdery soft material from Afghanistan!

The detailed study, *A Study of Precision Cleaning Methods for All Fiber Optic Connections,* with more than 60 additional pages and 100+ photographs of the actual evaluations is also available: Amazon and www.createspace.com/5271116.

Additional copies of this book are available: Amazon and on www.createspace.com/5173068

Pease inquire about special e-Book versions and printed copies at discounted pricing for qualified organizations and training associations.

+770-971-8100
edforrest@live.com

Q: WHY TEST WITH SO MANY DIFFERENT CONTAMINATION TYPES?

A:

1.) Because fiber optic networks are deployed in many ambient conditions in all parts of the world.

2.) If the fiber optic end face is not cleaned...properly...it will not test or transmit to design capacity.

3.) "Worst Case" leads to
                              "Best Practice"

Deployments can be a harsh, diverse environment with myriad debris and contamination types

Sporting and Broadcast

Security. Intelligent Traffic Control

Inter-company communications

FTTx

Data Centers

DOD

Municipal Infrastructure

Commercial Aviation

Airborne Debris

Condensation Contamination

Cellular Tower to MTSO

Healthcare

Evaluating "worst case" to "best practice".

How to create your own internal standard.

It may be that you want to create an "internal standard" based on your "Applications Specific" requirements. We will help you with this and the data will assist all of us.

Considered in this evaluation are
<u>three types</u> of contamination.   Select debris
based on what you might experience.

I.)

DRY DEBRIS:

a.) Arizona Road Dust

b.) Desert Dust from
   Afghanistan

a.)

b.)

<u>DID YOU KNOW?</u>

Since "dry debris" has *height* as well as *diameter*,
if the diameter of the debris passes a specific test
or standard, the height that can create a stand off
between the connections or damage the surface
at time of mating the connections.

II.)

Considered in this evaluation are <u>three types</u> of fluidic contamination

Fluidic Contamination:

c.) 10-40 Engine Oil

d.) Hand Lotion

c.) Vegetable Oil

DID YOU KNOW?

Fluidic Contamination is the most troublesome simply because it can move from area to area within the connector 'geometry'. It is one thing to 'view' a clear and clean Zone within the field of view...and another when fluids outside the 'field of view' transfer at time of mating the connections and post-cleaning.

HOW IS THE CONNECTION CLEANED TO "BEST PRACTICE"?

**Considered in this evaluation are
<u>three types</u> of contamination**

SINCE FIBER OPTIC DEPLOYMENTS ARE 'EVERYWHERE',
REMOVING THE WIDEST RANGE OF DEBRIS IS CONSIDERED
"BEST PRACTICE".

THE THIRD TYPE OF CONTAMINATION
MAY BE MOST COMMON:

III.)

<u>DRY DEBRIS AND FLUIDIC</u>
<u>CONTAMINATION</u>
<u>"IN COMBINATION".</u>

The possibilities are nearly infinite.

Establishing Internal Standards:

Suggested Test Protocols:

Use each of the cleaning device to clean different types of Dry Debris debris, Fluidic Contamination and Combinations of both .

Evaluate each cleaning device or procedure one time. If there is no appreciable removal, a second attempt will be made. Never use a cleaning surface more than once.

The intent is NOT to compare one against the other, but rather, suggest and indicate which product and process could work best for your specific application.

Each surface is cleaned to "pristine" between cleaning tests. The number of cleanings required for this was not published in this evaluation. The complete evaluation will be published at a future date.

Establishing Internal Standards:

Suggested Test Protocols:

Protocols for contaminating the end face is derived from 2006 instructions by CICSO® and evaluation of ten contaminants. [2]

- For "dry" and "fluidic" type comparisons, each device is lightly touched into the contamination.

  - For "combination types" devices are placed in "fluidic" (first) to attract contaminant and (second) in "dry".

    - Probes or swab tools are used to transfer contaminant to the SFP Transducer stub or a backplane connection through the alignment sleeve.

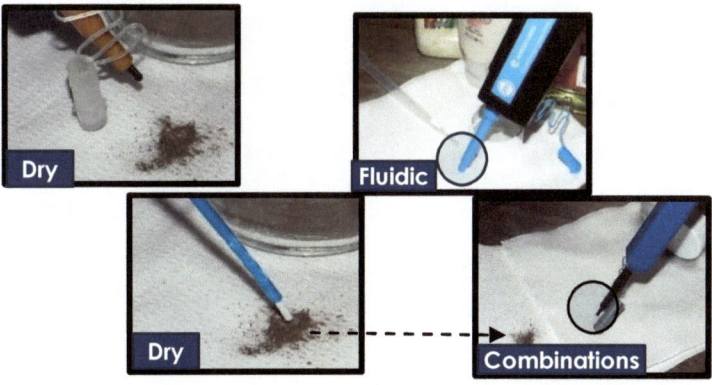

[2] Please request copy of evaluation

Establishing Internal Standards:
Suggested Goals and Protocols of the Experiments

▶ **First Time Cleaning of "End Face" (or "Stub") and jumper end face surface**

▶ **Complete documentation**

1.) **Create 'straight up, first-time tests', no retakes.**

2.) **Likely not all debris or contamination will be removed.   Your final results may be enhanced with a larger statistical sampling.**

3.) **Evaluate the relative ability of each process and specific product.**

Each surface may be cleaned in this manner:

1.) Dry debris types may use one or more swab tools.

- The number of passes are noted on each test sheet and are available on request.

2.) If a solvent it used with a swab tool, one additional swab tool may have been used to after-dry.

- There was one 'wet-to-dry' pass on a cleaning platform. A "pass" is defined by the size of the cleaning surface and may be 1-3 cleaning strokes.

3.) Fluidic contamination may be removed using one moistened swab, followed with a dry swab tool.

- Fluidic Contamination is removed by passing the end face over the cleaning platform.

44

Establishing Internal Standards:
What is the actual process?

1.) Select the debris or contamination you want to "standardize". Try to obtain sufficient quantities not only for this test period, but also for others in the future.

Remember, you will not need "gallons" or "pounds", but sufficient quantity to have for 5-10 years. "Future-Proof".

2.) Select your cleaning device. What you use now may not be what you will use next year or on the next installation. The only thing that may be the same is the debris.

3.) Purchase or rent a video scope. I prefer one with a wide "field of view". The one you have is fine!

4.) Test ten (10) times. If you can obtain 10 out of 10 cleaning events with 100% removal you have a "standard". That is what I want: 10 for 10.

However, you may decide that 9 out of 10 with "almost clean" is acceptable. You decide; you know the job; you know what the client expects.

5.) How to contaminate the end face repeatedly is suggested in this book on pages 42 and 48. This is done by 'contaminating' a clean wiper and then placing an end face in this debris.

This is a little art and science combined. Don't skew your own results" be fair and unbiased. If you need help...I am here!

Author's Technical Notes: In 2005 ITW Chemtronics® conducted a study of what was knows as the Cisco® Series of soils. The criteria was 100% clean, tested 10 times. There were some very unusual debris and contamination in this requirement.

45

Establishing Internal Standards:

Part-II The Evaluations

The "goal" is to provide a *sense* of which product and process will return a first time cleaning result.

Of course, there is both "reality" and "luck" in play here. Conclusive evaluations require many more attempts.

If you have questions, or *challenges*, please advise! Submit your specific requirements for further evaluation.

A 'guide-post' study with more than 100 pictures, using the following products is available at www.creativespace.com/5120367

Considering Probe Tools

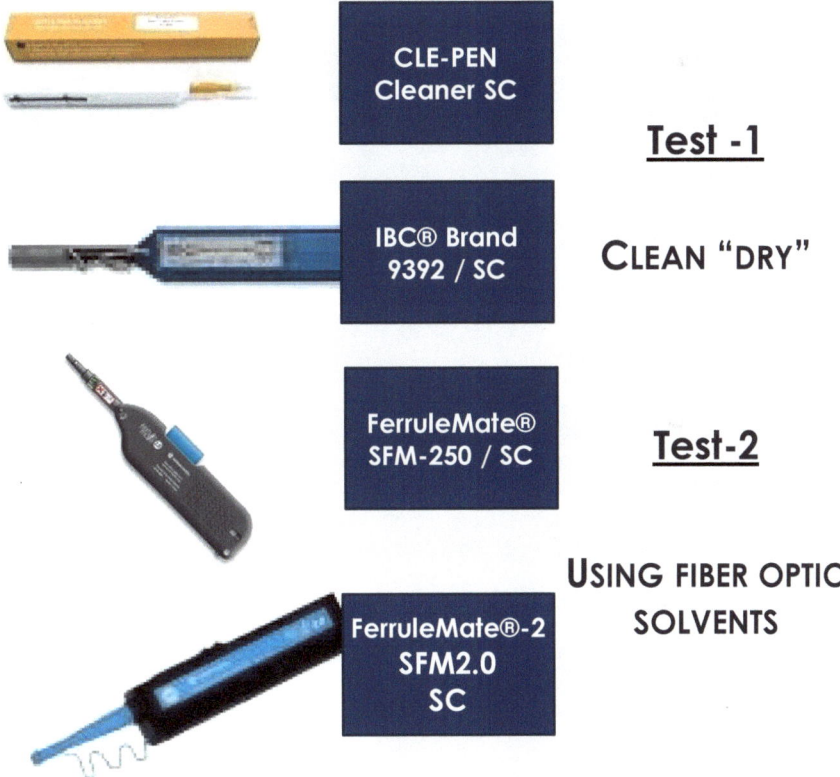

CLE-PEN Cleaner SC

Test -1

IBC® Brand 9392 / SC

CLEAN "DRY"

FerruleMate® SFM-250 / SC

Test-2

USING FIBER OPTIC SOLVENTS

FerruleMate®-2 SFM2.0 SC

| FerruleMate®-2 SFM2.0-250 SC | FerruleMate® SFM-250 / SC | IBC® Brand 9392 / SC | CLE-PEN Cleaner SC |

DIFFERENCES AND SIMILARITIES:

Each tool is designed for convenience with semi-automatic advance of a (WIDE) "TAPE" or (NARROW) "THREAD" to clean the end face.

The two devices on the -left- use "tapes" that clean a wider field-of-view.

The two devices on the -right- use "threads" that may not clean as much of the surface.

Each can be highly effective with used properly in conjunction with 100% video inspection.

1.) Removing Dry Debris using probe tools (image 1) **without a solvent.**

2.) When a probe tool is moistened it will be as shown (image-2) **with premium non-IPA fiber optic cleaners.**

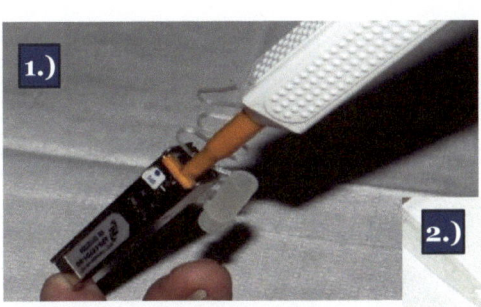

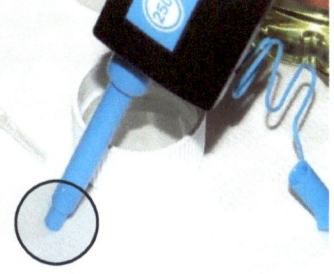

Insert the Probe Tool into the swab tool into the backplane/SFP type.

Activate per manufacturer's instructions and video inspect

If "dry cleaning" does not remove the contamination, lightly moisten probe tool in fiber optic grade precision cleaner

49

Considering Swab Tools

Sticklers® S25 2.5mm CleanStixx™

ITW Chemtronics® 48042F 2.5mm Short Handle

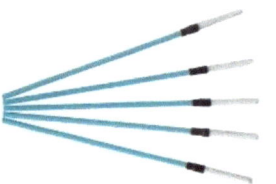

NTT-AT 2.5mm Fiber Optic Swab

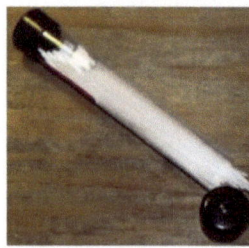

ITW Chemtronics® 48062F TorqueTube™ 2.5mm

Test -1

Each clean "dry"

Test-2

Each clean using fiber optic solvent as indicated if the surface is not cleaned the first time*

* The manufacturer of the NTT® swab does not recommend use of a solvent

50

How is the connection cleaned to "best practice" using a swab tool?

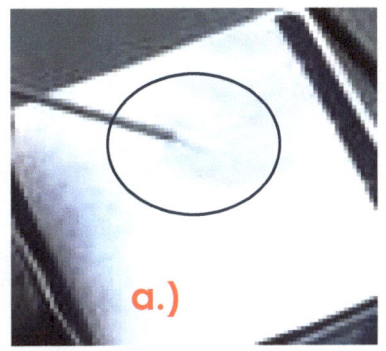

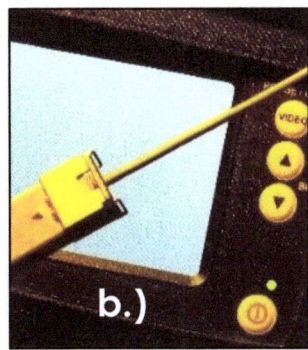

a.) Lightly moisten the swan head by placing it in a "spot" of precision fiber optic cleaner

- A swab tool rotated in the alignment sleeve may also clean the alignment sleeve prior to being pressed against the stub or back plane end face.

  b.) Rotate the handle...between thumb and forefinger, two or three times. Video Inspect; dry if necessary.

- Swab tools are used not only for 'back plane end face/stub' but also to clean contaminated inter-surfaces .

DID YOU KNOW?

All swab tools in this evaluation are all constructed of superior materials including premium microfiber, sintered polyester, and cleanroom grade medical foam.

Considering Cleaning Platforms™ and Reel Cleaners

THE "PROS" AND "CONS"

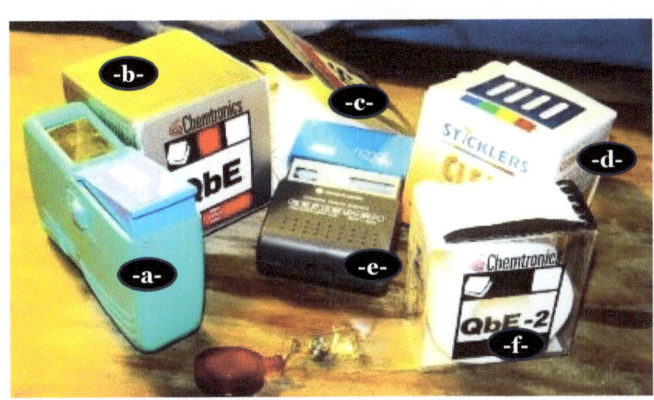

Reel Cleaners such as
a- CleTop-S® are
convenient.

Cleaning Platforms(tm)
b- Chemtronics® QbE®,
c- SqR™,
d- Sticklers®
CleanWipes™,
e- Seikoh Giken
HandiMate®,
f- QbE-2™

play important roles in
"precision cleaning"
beyond "convenience"

As concluded with the swab tools™ and probe devices, the actual
amount of cleaning surfaces influences the end result. ProbeTools™
are fast and convenient. Swab Tools can be more economical, but
difficult to use. Swab tools are valuable to clean surfaces other than
an end face.

Cleaning Platforms® have significant advantages to ProbeTools™
and SwabTools™ in that the larger cleaning surface moves debris
and contamination *away from the initial point of contact.*

SwabTools™ and Probes rotate and recirculate the contamination
which means repeated cleaning may be necessary and 100% video
inspection mandatory.

Reel cleaners are convenient but have a smaller cleaning surface
not unlike swabs and probes.

Is there a "real difference" between 'cleaning platforms' and 'reel cleaners'?

1-There are numerous 'cleaning platform' designs. They are characterized by a larger cleaning surface: while they are "thicker" than most reel cleaners, both are about the same cubic inch measurement.

2.) One cleaning platform has a patented feature: a compliant platen surface on which the special wiping material lays flat. The compliant surface compensates for APC geometry as well as provides a softer surface which can keep grit from being ground into the end face.

Other platforms have a hard cardboard surface or 'suspend' the wiping material. Each cleaning platform is a completely new tool.

3.) In addition to the original CleTop®, there are several designs based on the   original NTT® patent.

Reel Cleaners are thinner and may be placed in a pocket or vest pouch. Their small surface is not conducive for "wet-to-dry" cleaning. Since the surface is small (1"/2.5cm) fiber optic cleaners tend to leach and reduce the drying surface area. This is especially problematic for aqueous cleaners which require a 'drying runway'.

4.) Reel Cleaners have hard platens produced typically of a natural or synthetic rubber like material. These platens are not replaceable.

Consider all factors when selecting one of these tools.

How is the connection cleaned
on a cleaning platform or reel cleaner?

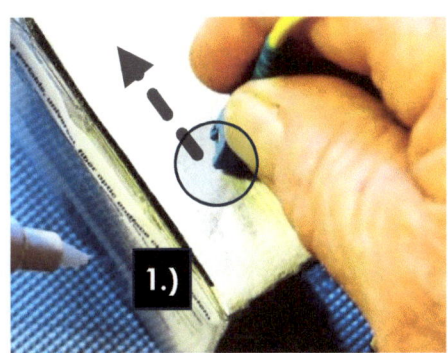

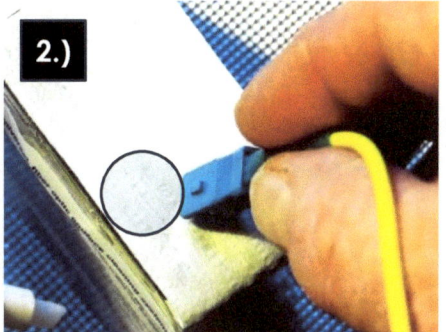

1-Lightly moisten the cleaning platform.

Glide the end face surface away from the initial point of contact to adequately dry as well as not recontaminate.

2-Make a conscious effort to position the end face at a perpendicular to the cleaning platform.

3-Compensate for APC by "look & feel". If there is a 'drag' then you are likely 'off angle'.

REMEMBER: Precision cleaning is a combination of "science" and craftsmanship. In the science of cleaning...just about anything...contamination is drawn to moisture.

Dry methods are typically used to "mop" fluidic contamination and are often not effective on dry type debris.

When using a cleaning platform the length of the cleaning stroke provides better cleaning. The length also "feeds-back" to you: if the end face does not "glide" then you are likely not "on angle" and may not clean completely. Swab tools and probes cannot provide this valuable sense of "finding the angle".

Techniques to properly clean:
The "Figure-8" Motion is a "Polishing Action"

Fusion Splice Prep is not End Face Cleaning.

Some propose the cleaning action for an end face is the "Figure-8".
That motion works well to finalize a spliced connection.

1. The starting point of the motion: the end face surface is contaminated.

- From the starting point the end face is drawn a, b, c, d, e, f, g, h.

To a point where the motion may start again and recontaminate.

2. Is the "critical juncture".

- At the beginning of the "Figure-8" motion, debris is deposited along the pathway.
- As the motion continues, the end face passes and re-traces at -2- where it may be recontaminated.

✓ The fiber optic "cleaning motion" is a *straight line action* that moves debris and contamination *away from the initial point of contact.*

56

How to select a fiber optic cleaner: Worker safety and environmental concerns are always top criteria. Be sure you read and have MSDS explained to you: ask the provider. Merely *having* an MSDS is not the point OSHA intended to educate you...it's important to understand the document itself.

This range of chemicals represents the 2015 list of best choices. These are general chemical families and each company has a 'trade name' for their product. Ask your supplier *"what's in this stuff?"*.

To "future-proof" you also have to self-educate.

HFE 7100 w/IPA/CZ®	Precision Hydrocarbon Formulations	Aqueous (Glycol Ethers)

▸ Advantages
  ▸ Numerous formulations
    ▸ Very good cleaning
  ▸ Convenience containers
    ▸ Easy Ship
    ▸ Aerosol
    ▸ Non-flammable
 	
▸ Disadvantages
  ▸ Ultra-Fast Evaporating
    ▸ Can leave residues from some soils.
  ▸ Highest cost
	▸ Advantages
	  ▸ Numerous formulations
	    ▸ Wide range cleaning
	    ▸ Check
	  ▸ Convenience Containers
	  ▸ Low cost
	
	▸ Disadvantages
	  ▸ DOT regulated shipping
	    ▸ As with IPA
	  ▸ Flammable
	    ▸ As with IPA
		▸ Advantages
		  ▸ Newest formulations
		    ▸ Wide range cleaning
		    ▸ Check
		  ▸ Growing demand in many segments
		  ▸ Convenience Containers
		    ▸ Easy Ship
		  ▸ Lowest cost
		
		▸ Disadvantages
		  ▸ Must dry with a 'wet-to-dry' step

Test and compare. Try not to let 'aroma' be a criteria unless there are specific and individual reactions. A chemical choice is based on many factors and the actual cleaning ability is the foremost consideration. Demand training from your supplier: it may be no charge or fee based.

How to select a fiber optic wiping material

Many types of cloth have been created since the beginning of time. Earliest came from animal skins which were superseded by weaving. Now, there are synthetic animal skins and woven cloth of natural and synthetic materials.

There is also a new-generation of material that weaves or 'hydro-entangles' synthetic and natural materials. Selection of a strong material that does not shred or leave a surface residue is a critical concern to the precision fiber optic cleaning process.

100% Cellulose (paper)	Polyester / Microfiber	Hydroentangled polyester/cellulose

▸ Advantages
  ▸ Absorptive
  ▸ Convenience package
  ▸ Readily Available
  ▸ Lowest Cost
 	
▸ Disadvantages
  ▸ Low sheer strength
    ▸ Tears, shreds with Lint residues
  ▸ Used for many applications
    ▸ Readily available in the supply chain
	▸ Advantages
	  ▸ Absorptive on most debris
	  ▸ High sheer strength
	    ▸ Low Linting
	  ▸ Often used in probes and other cleaning devices
	    ▸ Readily available
	  ▸ Moderate cost
	
	▸ Disadvantages
	  ▸ Can create a static charge
		▸ Advantages
		  ▸ Highly absorptive
		  ▸ High sheer strength
		  ▸ Often used in cleaning platforms
		    ▸ Readily available
		  ▸ Moderate cost
		
		▸ Disadvantages
		  ▸ Many choices and not all perform to the same level.

Test and compare. 100% cellulose (paper) and 100% cotton are not recommended to precision clean a fiber optic connection. Likewise, any treated material, possibly with an electro-static discharge (ESD) compound is not suggested as there can be a residue especially when a solvent is used.

58

What Does Not Work and Why Not

1.) IPA in all forms

➢ Chemically not suitable to remove a wide range of debris and contamination. Good for perspiration and salts; not so much for other types.

➢ Hygroscopic: IPA attracts moisture and deteriorates cleaning ability

2.) Pumps and Squeeze Bottles

➢ Always maintain the integrity of the cleaner. These devices bring deteriorating moisture into the container.

3.) Pre-Moistened lens grade optical tissues…may have surfactants.

➢ Superior for a microscope or eyeglasses: *not fiber optics*

4.) Paper Products. Cotton.

➢ Lint residues not conducive to high speed and bandwidth

5.) "Canned Air"…compressed gas dusters

➢ Even high velocity dusters cannot remove 'surface bonded" debris*,

➢ Some flammable, others not plastic safe.

➢ May be useful to reclaim water or storm damaged connections.

59

* White Paper on Request

What Works Best and Why

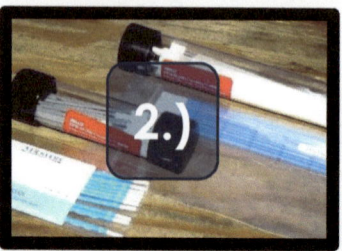

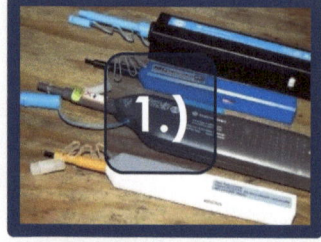

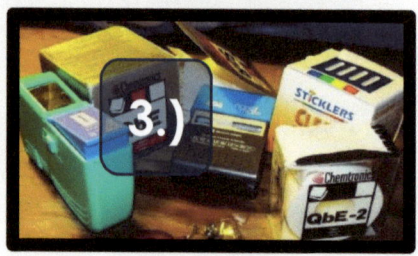

❑ **Precision Cleaning a Fiber Optic Connection is an Applications Specific Process.**

❑ **Each of these products can work better when lightly moistened with a non-IPA Fiber Optic Cleaner**

➤ Not all of these products work the same way to return the same results.   Select and use the ones that are best for the specific application. Challenge the manufacturer to train and demonstrate before you decide.

➤ Not all fiber optic cleaners have the same performance characteristics: select the right one based on ability to clean anticipated dry debris, fluidic contamination or combination of types.

BEHIND THE SCENES:

Desert dust from Afghanistan is significantly finer than Arizona road Dust or Huntington Beach Sand! In the image (left) residual contamination is surface bonded to the plastic substrate. "Surface Bonded" debris or contamination is more difficult to remove than that laying on top of a surface: it is held in a monomolecular layer. Residual contamination may be "dry" or "fluidic" and be a result of an inadequate cleaning process. The debris to the left must be 'touched' with a moistened cleaning surface to remove it. Interested in learning more? Ask for White Paper information on surface bonding and residual contamination.

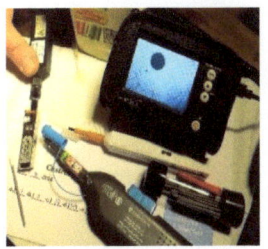

To remove some of the more complex contamination such as 10w-40 engine oil, several devices were used. The intent of this study is not to declare a 'winner', but rather demonstrate that cleaning is applications specific.

That simply means that one device may not be as good as an other...but not that one is better than the other all the time! Match the cleaning tool with the anticipated debris or contamination. "One size may NOT fit all" cleaning needs!

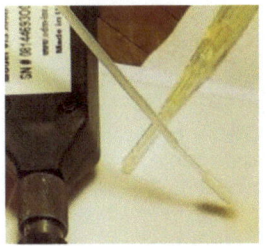

Make an effort (evenly as possible) to contaminate the SFP stub or 2.5mm jumper. This image depicts 10w-40 engine oil siphoned to a clean pipette and then deposited on a clean cellulose-polyester non-woven wiper.

A clean swab tool was placed in the spot; the swab tip contaminated the stub or the 2.5mm end face was touched to the wiper. Same was done for "dry debris" as well as "fluidic contaminants" and "combinations".

See something you don't "like"? Please request a re-test!

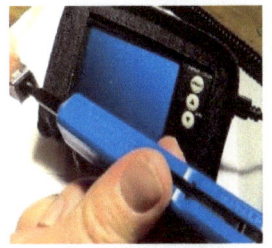

Each of the probe tools required an "acquired touch" to place the cleaning tip into the SFP connection. That means that the nose on the probe did not intuitively align as well as a swab which "fit".

Be certain the cleaning tip is placed into the SFP transducer...one some probes it might "miss" the alignment and surely not clean. Swab heads were easier to place, but required more effort and were less convenient. (See Slide 5)

61

Behind the Scenes

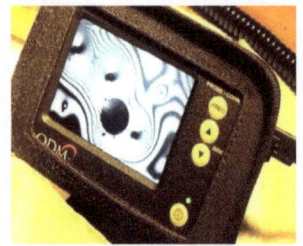

The value of a 'high resolution" video inspection device is readily apparent. Here is not only a view of "hybrid" or "combination contamination", but also its' diameter as well as the height can be discerned. Usually an interferometer is required to discern height: the resolution of this video scope emulates 'height'.

Between each sitting clean the Zone-5 of the SC Jumper to assure debris or contamination would not leach from the vertical ferrule to the horizontal end face.

The side wall of the SPC™ transceiver was only 'touched' by the foam swabs. All other devices were deemed too narrow to clean the internal side wall.

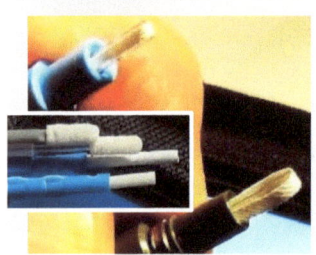

The size of the cleaning surface, in relation to the type of debris or contamination, is relative to the specific cleaning ability of any given product. On the top and bottom left are tapes from probe tools; in the center are swab heads. These larger surfaces than "threads" may equate to better cleaning ability.

In general terms, the larger the cleaning surface the *better it will clean and absorb.*

The "go-to" means of cleaning the SFC™ Transceiver was primarily the SFM-250 FerruleMate. The wider tape clearly removed more debris and contamination with one to two passes. As the testing progressed, FerruleMate-2 was also used. Moistened ClePen® and IBC® Tools also were used from time to time during these considerations. Limited quantities of some swab tools made their use for 'in between tests" impractical.

62

1.) ___T___F   The first thing a technician should do is observe the ambient environment and get an idea of the type of debris or contamination that may be present.

2.) ___T___F: 99.9% isopropyl alcohol (IPA) is the best cleaner because it removes the widest range of debris and contamination all the time.

3.) ___T___F: Cotton and paper are just as good as microfibers and non-woven cellulose/polyester blends to hold and absorb debris and contamination.

4.) ___T___F: "Field of View" refers to the area seen by most video inspection

5.) ___T___F: Fluidic contamination stays in place and dry debris moves

6.) ___T___F: One problem with swabs is that foam is not as good as other materials to absorb

7.) ___T___F: All fiber optic cleaning solvents are the same; they are simply packaged differently

8.) ___T___F: If I do not have a video scope, cleaning multiple times works just fine

9.) ___T___T: A major advantage of all "probe tools" is their convenience

10.) ___T___F: When using a cleaning platform, hold the end face at 90 degree perpendicular to the cleaning surface for best result.

11.) ___T___F: The "best practice" fiber optic cleaning technique is a "Figure-8" motion because it is fast and repetitive.

12.) ___T___F: "Wet to Dry" cleaning does not require video inspection.

13.) ___T___F: The straight line cleaning action moves dry debris and fluidic contamination away from the initial point of contact.

14.) ___T___F: Fiber optic standards, such as IEC 61300-3-35, are updated every 5 years with annual bulletins and updates.

15.) ___T___F: Many of the same cleaning techniques and products typically used for fusion splice prep are appropriate for end face cleaning.

ANSWERS: TEST YOUR KNOWLEDGE:

1.) ___T   The first thing a technician should do is observe the ambient environment and get an idea of the type of debris or contamination that may be present.

2.) ___F: *IPA sterilizes and is not an effective cleaner on wide-range soils.*

3.) ___F: *While cotton and paper may absorb, they are also leave lint and paper shards.*

4.) ___F: *Most scopes used magnification and this limits 'field-of-view".*

5.) ___F: *It's the other way 'round!*

6.) ___F: *Foam absorbs well; as with anything make sure the foam is designed for this application.*

7.) ___F: *There are excellent fiber optic cleaners and others not so much.*

8.) ___F: *The "combination or hybrid method" works best when blind cleaning. Cleaning multiple times, especially if the debris is hard material, can damage the surface.*

9.) ___T: A major advantage of all "probe tools" is their convenience

10.) ___T: When using a cleaning platform, hold the end face at 90 degree perpendicular to the cleaning surface for best result.

11.) ___F: *The "Figure-8 Motion" is a polishing action associated with repair or replacement of the connector end.*

12.) ___F: *Every connection, no matter which technique, should be video inspected. However, the "hybrid" or "combination" technique works best because the amount of solvent is limited and the surface is dried.*

13.) ___T: The detail of this procedure is common sense. The fiber optic surface is so small it is easy to contaminate and re-contaminate.

14.) ___F: *As of 2015, fiber optic standards for precision cleaning and inspection are updated every ten years.*

15.) ___F: *Fusion splice prep is a completely different application than end face cleaning. For fusion splice, the side and length of the fiber is cleaned. For this, IPA works well enough. Cross-use of IPA to end face cleaning can result in inferior results.*

Additional Notes and Comments:
Establishing Internal Standards

→Contamination of each surface is accomplished at the stub of the transceiver by lightly placing a 2.5mm swab tool into the debris and touching the surface.

→Dry debris is distributed in as random a fashion as possible.

→Fluidic Contamination was distributed by lightly touching the swab tool head into the moisture and then lightly touching the end face

→Contamination of the 'jumper side' is accomplished by lightly touching the end face into various debris and then touching that surface into a clean area of a non-woven hydro-entangled wiper surface. Dry debris was distributed in as random a fashion as possible.

→Fluidic Contamination is distributed by lightly touching the end face into the moisture and/or subsequently touching the end face

→ Observation: it does not "require" a significant quantity of any debris or contaminate to foul a fiber optic surface.

For some, it is difficult to imagine how a topic as mundane as cleaning a fiber optic connection can influence an entire Industry. After all, who has not (and may still) cleaned the surface on a clean t-shirt, or as one technician (seriously) advised me: "*under the collar of a shirt is better*"!

If you are on a production line, how you are taught to clean the surface not only influences the final product, but also its' installation. If you are an installer, and you are not able to establish service, you may struggle cleaning the surface four or five times, only to return a component to the producer...only to have a warranty repair embarrassingly rejected because the fiber optic surface was contaminated and not properly cleaned! For some reason, within all the brilliance of fiber optic design and deployment, we seem to have lost sight of an important fundamental: *if the connection is not properly cleaned is will not transmit to design standard and surely will not test with accuracy.* We seem to know more about cleaning a floor than precision cleaning a fiber optic device! Surely, there are as many products!

The Telecommunications Industry is in continual evolution. Not too many years ago cables were laid from the back of horse and buggy: now fiber optic lines are deployed trans-continental, trans-oceanic, and to the home or office desk. The end user expects reliability and capacity: *what is deployed now, was once considered theoretical only a few years ago.* New connections are always entering the market: just a few years ago 'expanded beam' was considered 'no clean' until the realization that fluidic contamination can transfer! The same may happen with the humble Jumper-Side SC connection as icing from a morning donut transfers to the backplane and may foul the adapter sleeve during insertion! "Worst case leads to Best Practice".

The results of this study indicate that just about any cleaning product can be made to work! However, without 100% video-inspection any cleaning procedure is truly a wish and a hope. It is a hopeful result this study helped others understand these important tenets: 1.) there are many types of potential debris and contamination to consider, 2.) not all cleaning products and associated procedures return the same results, 3.) 'first time cleaning' of some debris or contamination is possible. 4.) "blind cleaning" is far more common than many in the Industry feel comfortable accepting, 5.) precision cleaning each time a connection is 'made' is fundamental best practice, and 6.) video inspection is *how to measure cleanliness* and not (common place) reliance on a light source and power meter.

Challenge all suppliers of cleaning products by asking for performance comparisons and possibly creating your own internal standards. Don't blindly accept a product claim any more than you would blindly-clean in a leap of faith!

All the best,

Ed Forrest

Bringing Ideas Together™

RMS (RaceMarketingServices™)
est: 1974

by: Edward J. Forrest, Jr. {Ed}
Founder and Inventor
1800 Basswood Court
Marietta, GA. 30066-2929 USA
+770-971-8100

edforrest@live.com
www.fiberopticprecisioncleaning.com
www.cedarkeyinstitute.com

2015 All rights reserved

ABOUT THE AUTHOR:

Ed Forrest has been actively involved in specification and applications engineering of various precision cleaning applications for more than 25 years. Previously employed at ITW Chemtronics®, retired in July-2014, he was schooled to analyze precision and gross cleaning applications in a wide range of applications. In 2001 he began development of a program that resulted in formal approvals at all major telecommunications providers.

He has seven patents specifically in the areas or fiber optic precision cleaning with six products in production with marketing credits that include branding, training, and publication of materials. He innovated a chemical mid-span break-in for ribbon fiber. He has other patents pending.

He is active on fiber optic standards committees and is considered a SME in the study of fiber optic cleaning and inspection. His work is based on field experiences and the needs of designers, crafts persons and production line workers.

His practical thesis of "Five Zone Cleaning" is a look forward to the times when high speed and capacity of fiber optic transmission (even more) will be impacted by a contaminated or improperly cleaned connections. He has uniquely researched inspection of the 4th and 5th Zone and the influences of various debris and contamination as it is positioned on these areas of the connector.

He worked as an Electronics Manufacturer's Representative throughout the 1970's. He actively participated in the early introduction of some of the most fundamental electronic products in the changeover from analogue to solid state. These included solid state components, consumer products including the first hand-held calculators, esoteric high fidelity, test equipment, games and other electronic products considered 'cornerstones' of the contemporary marketplace. He has production credits in that Industry

He worked in a then-developing market segment in the Home Furnishings Industry. By coordinating North American and International Development, using an effective agency in Denmark he was able to work throughout Europe prior to the time of the EU. In coordination with C.ITOH (est-1860), he traveled and developed a Japanese market long before current interest in the important nations of The Pacific Rim. He initiated promotional activity in conjunction with USA Embassies, individual USA states resulting in active trade in Denmark, Sweden, Finland, Italy, Germany. Great Britain, nations in The Middle East and South Africa. He has production credits in that industry.

Early career as a Technical Representative, in Union Carbide Corporation's Automotive Consumer Products group, career-forming experiences include introduction of Prestone® AntiFreeze as a Summer Coolant in a one year NASCAR race test and associated promotions, as well as, an innovative time with Standard Oil of Ohio® as SOHIO® introduced "self-service fueling" to the market. He competed in the market when brands like STP® and Wynn's® dominated consumer interest.

He is an active photographer, enjoys study of the ancients, and a hobbyist collector of esoteric high-fidelity. As a life-long SCCA member he competed "wheel-to-wheel" in more than 200 events at SCCA's Club Racing level in cars he designed...'with a little help from his friends'. Married with a fascination for Weimaraners, he and his wife are often 'at the edge' with three lovely specimens. They have travelled extensively through the USA and Europe.

Notes:

Je Suis
Charlie

www.ingramcontent.com/pod-product-compliance
Lightning Source LLC
Chambersburg PA
CBHW040840180526
45159CB00001B/257